OUR SOLA

GIANT PLANETS

Robin Kerrod

Belitha Press

First published in Great Britain in 2000 by

(⑤) Belitha Press
A member of **Chrysalis** Books plc
64 Brewery Road, London N7 9NT

Paperback edition first published in 2003
Copyright © Belitha Press Limited 2000
Text by Robin Kerrod

Managing editor: Veronica Ross
Editor: Jinny Johnson
Designers: Caroline Grimshaw, Jamie Asher
Illustrator: David Atkinson
Consultant: Doug Millard
Picture researcher: Diana Morris

ISBN 1 84138 060 1 (hb)
ISBN 1 84138 754 1 (pb)

British Library Cataloguing in Publication Data for this
book is available from the British Library.

Printed in Hong Kong

10 9 8 7 6 5 4 3 2 1 (hb)
10 9 8 7 6 5 4 3 2 1 (pb)

Picture credits

Robin Kerrod/Spacecharts: 12tc, 30b, 33t.
NASA/Spacecharts: 1, 3, 4-5, 6c, 7, 8t, 10, 11, 12tl,
.2b, 13, 14c, 15b, 17, 18, 19, 20, 21, 22, 23, 26, 27, 29, 30t,
31, 33b, 35, 36t, 37, 38, 39, 40, 41, 44, 46-7.
Science & Society P.L.: 6t.
Tempio Malatestiano, Rimini/Bridgeman Art Library: 14b, 28b.

Some of the more unfamiliar words used in this book are
explained in the Glossary on pages 46 and 47.

CONTENTS

INTRODUCTION

Our home planet, the Earth, seems a big place to us. And it is bigger than its neighbours in space, the near planets Mercury, Venus and Mars. But, compared with the planets Jupiter and Saturn, the Earth is a dwarf.

Jupiter and Saturn are truly gigantic and are well called the giant planets. Jupiter is biggest, more than ten times bigger across than the Earth. It could swallow more than 1300 bodies the size of the Earth. The only body in the Solar System bigger than Jupiter is the Sun.

Jupiter and Saturn lie beyond Mars in the Solar System, going away from the Sun. They lie much farther away from us than the near planets.

Saturn is the most distant planet that can be seen with the naked eye. No one knew there were any more distant planets until 1781, when Uranus was discovered.

The giant planets are very different kinds of worlds from the tiny Earth and its even smaller neighbours. The Earth and the near planets are solid bodies made up mainly of rock. Jupiter and Saturn are made up mainly of gas. They have no solid surface at all.

The giant planets are different from Earth in yet another way. They are like miniature solar systems with many satellites, or moons, circling around them. Saturn alone has more than 20 moons.

DISTANT WORLDS

Jupiter and Saturn lie far beyond our neighbour Mars.

At their brightest, Mars and Jupiter appear about as bright as each other in the night sky. So we might be tempted to think that they are about the same distance away. But nothing could be further from the truth.

Say we could ride to the planets in a space shuttle. A trip to Mars when it was closest to us would take about three months. Mars is one of our planetary neighbours, sometimes coming as close as 56 million kilometres.

A trip to Jupiter when it was closest would take more than two years. The planet never comes closer to the Earth than about 600 million kilometres. To continue on to Saturn would extend our journey time to nearly six years. Even at its closest to Earth, Saturn lies over one-and-a-quarter billion kilometres away.

▷ **Only the planets out to Saturn are shown in this orrery of 1733.**

▽ **Saturn from a *Voyager* space probe.**

Sun   Mars   Jupiter   Saturn

It is incredible that we can spot Saturn with the naked eye at such a distance. But it is the most distant planet that we can see with just our eyes. We need a telescope to see the three planets further out - Uranus, Neptune and Pluto.

How the giants formed

The Sun and all the planets formed about 4600 million years ago. The planets were born out of a disc of gas, dust and chunks of rock. They grew in size as the bits of

▽ **Jupiter and Saturn lie far from the Sun.**

Uranus

△ **Neptune and its large moon Triton can only be seen through telescopes.**

matter kept hitting one another and sticking together. The inner part of the disc was hot. There, four planets – Mercury, Venus, Earth and Mars – formed out of rocky lumps. But the gas in this region, mainly hydrogen, did not stay there. The heat and other rays from the Sun forced the gas into the outer part of the disc. There it was much colder. The gas began gathering around the small rocky lumps that had formed there. There were enormous amounts of gas, and the small lumps soon grew into huge bodies of gas. These became Jupiter and Saturn, the giant planets we know today.

GAS BALLS

Jupiter and Saturn are great balls of gas and liquid gas.

Jupiter is a huge planet, with a volume more than 1300 times that of the Earth. But it has only about 300 times the Earth's mass. In other words, it is much lighter for its size than the Earth is – it has a much lower density. This is how we know that it must be very different in structure from the Earth.

Jupiter is too light for its size to be made of solid rock like the Earth is. It has to be made up of much lighter materials, such as gas. Studies from the Earth and by space probes have shown that Jupiter is made up mainly of the two lightest gases there are – hydrogen and helium.

△ The shuttle's main engines use liquid hydrogen as fuel.

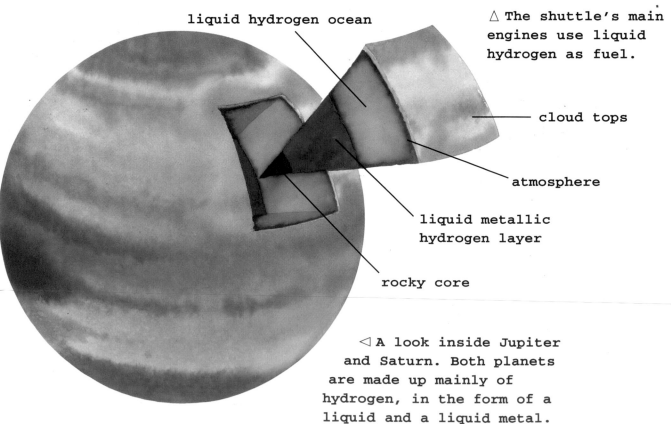

liquid hydrogen ocean

cloud tops

atmosphere

liquid metallic hydrogen layer

rocky core

◁ A look inside Jupiter and Saturn. Both planets are made up mainly of hydrogen, in the form of a liquid and a liquid metal.

Under the clouds

Astronomers think they know what Jupiter is like inside. The part of the planet we see in telescopes is the top of a deep atmosphere, containing mainly hydrogen and helium. This atmosphere is probably about 1000 kilometres thick. Underneath, there is not a solid surface, but a vast ocean covering the whole planet.

It is not an ocean of water as on Earth, but an ocean of liquid hydrogen. This is hydrogen gas that has been changed into liquid by the pressure of the deep atmosphere above it.

Metal and rock

The liquid hydrogen ocean is as much as 20 000 kilometres deep, or nearly twice the diameter of the Earth. At the bottom of this ocean, the pressure is unbelievably high and crushes the very atoms of hydrogen. It forces them to form a kind of liquid metal, rather like the liquid metal mercury found on Earth.

This layer of liquid metallic hydrogen is twice as deep as the liquid hydrogen ocean above it. Astronomers think that it sits on top of a ball of rock in the centre of the planet. This rocky core is probably half as big again as the Earth in diameter. And it must be very hot. Its temperature may be as high as 30 000°C. This is more than five times the temperature on the surface of the Sun.

Inside Saturn

Saturn appears to be a smaller version of Jupiter. It is also made up mainly of hydrogen and helium, and almost certainly · has a similar structure. Under a deep atmosphere lies a liquid hydrogen ocean, and under that a liquid metallic hydrogen layer. Its rocky core, though, is probably much smaller.

▽ **Jupiter and Saturn dwarf the Earth but are tiny compared with the Sun.**

Earth

Saturn

Sun

Jupiter

LOOKING AT GIANTS

Jupiter and Saturn look spectacular in a telescope.

Ancient peoples all over the world were familiar with Jupiter and Saturn, two of the bodies they called wandering stars, or planets. Of the two, Jupiter is by far the brightest. For much of the year, it shines brightly in the night sky like a beacon, outshining all the other stars by a wide margin.

The only objects brighter than Jupiter in the night sky are the Moon and Venus. Venus appears only fleetingly in twilight skies at dawn or sunset. Jupiter appears in dark night skies and may be visible all night long.

Mars on occasions rivals Jupiter in brightness, and is also seen in dark night skies. But it is easy to tell them apart. Jupiter shines a brilliant white, whereas Mars has a reddish hue. This is why Mars came to be called the Red Planet.

△ **Large telescopes show many details on Jupiter's colourful disc.**

◁ **Mars (left) and Jupiter outshine the stars in the evening sky.**

◁ **Galileo's sketch shows Saturn's 'ears'.**

Finding Saturn

At times Saturn can shine as brightly as the brightest stars, but it does not stand out like Jupiter does. It is often difficult to spot unless you know exactly where to look for it. Details about where in the heavens Saturn and the other planets are visible at any particular time can be found in astronomy magazines.

In binoculars and telescopes

When you look at Jupiter and Saturn in binoculars, you will see each one as a definite circle, or disc. This is in contrast to the stars, which always appear as tiny pinpoints of light in binoculars. Looking at Jupiter, you will also see tiny bright dots lined up on either side of the planet's disc. These are some of the

Saturn's 'ears'

The Italian astronomer Galileo first observed Jupiter and Saturn through a telescope. He spotted Jupiter's four large moons, now known as the Galilean moons. He also noticed that there was something peculiar about Saturn. He said it had 'strange appendages', which he thought might be large moons. What he had spotted were Saturn's rings.

four large moons of Jupiter. You can see them change position from night to night as they circle round the planet.

But it is in telescopes that Jupiter and Saturn begin to reveal their true beauty. Colourful bands show up on Jupiter's disc, together with all kinds of other markings. Saturn looks even more spectacular because of its magnificent rings.

◁ **Telescope views of Saturn reveal the beauty of its ring system.**

CLOSE ENCOUNTERS

close
encounters

Space probes have unravelled the mysteries of the distant giants.

In March 1972, a NASA space probe named *Pioneer 10* lifted off the launch pad at Cape Canaveral in Florida. It was heading for a rendezvous with Jupiter in December 1973, far away in the Solar System beyond the asteroid belt. *Pioneer* was an appropriate name for the probe because no spacecraft had ever made the journey before.

After its encounter, or meeting, with Jupiter, *Pioneer 10* sped off into space. By 1997, it had travelled more than 10 billion kilometres away and was approaching the edge of the Solar System.

An identical sister craft, *Pioneer 11*, followed *Pioneer 10* to Jupiter, reaching the planet in December 1974. It looped round Jupiter and then headed for Saturn, which it reached in September 1979.

The two *Pioneers* blazed a successful trail to the outer planets, taking the best pictures yet of Jupiter and Saturn. They sent back much new information, such as data about the powerful magnetism of the two planets. *Pioneer 11* also discovered new rings around Saturn and a new moon.

△ *Pioneer 10* carries a plaque (above) giving information for aliens.

◁ This *Pioneer 10* photo shows swirling currents around Jupiter's famous Great Red Spot.

Space Voyagers

In 1977, two more space probes set off from Earth to explore the giant planets. Named *Voyager 1* and *2*, they had better cameras, instruments and computers than the *Pioneers*. They were programmed to take close-up pictures, not only of the two giant planets but also of their moons.

During their encounters with Jupiter (February and July, 1979), the two *Voyagers* sent back many superb pictures. They discovered a ring and several new moons and they showed that Jupiter's four large moons were amazingly different from one another.

When they arrived at Saturn (November 1980 and August 1981), the *Voyagers* spotted storms in the planet's atmosphere and many new moons. They sent back spectacular views of Saturn's rings, showing that they are made up of thousands of little ringlets. After leaving Saturn, *Voyager 1* began heading out of the Solar System, but *Voyager 2*'s mission was far from over. It had two more planets to visit – Uranus and Neptune.

▷ *Voyager 2* is launched by a Titan-Centaur rocket on August 20, 1977.

Explorations continue

Further exploration of Jupiter and Saturn from space did not start until the 1990s. The Hubble Space Telescope, launched in 1990, began taking regular pictures to show how the two planets change as time goes by. The probe *Galileo* began investigating Jupiter and its large moons when it went into orbit around the planet in 1995. It also dropped a probe to report on conditions in Jupiter's atmosphere. The probe *Cassini* set off in 1997 bound for encounters with Saturn and its moon Titan.

△ The *Voyagers'* better cameras showed Jupiter's atmosphere much more clearly than ever before.

KING OF THE PLANETS

Jupiter has twice as much mass as all the other planets put together.

▽ Jupiter from a *Voyager* probe.

△ The symbol for Jupiter represents the god's lightning bolt.

▽ Jupiter is the fifth planet from the Sun, beyond the asteroid belt.

Ancient peoples believed Jupiter was a very important body because it always shines brightly when it appears in the dark night sky. The ancient Greeks called the planet Zeus after the most powerful god in their mythology. Zeus was king among the gods, the ruler of the heavens.

We call the planet Jupiter after the Roman name for this same god. Jove is another word for Jupiter, and astronomers often use the term Jovian to describe things relating to Jupiter, such as Jovian moons.

Jupiter shines so brightly in the night sky for two reasons. One, it is very big. With a diameter of some 142 200 kilometres, it is 11 times the size of the Earth. Two, Jupiter has a thick, cloudy atmosphere that reflects sunlight well. Like all the planets, it gives out no light of its own.

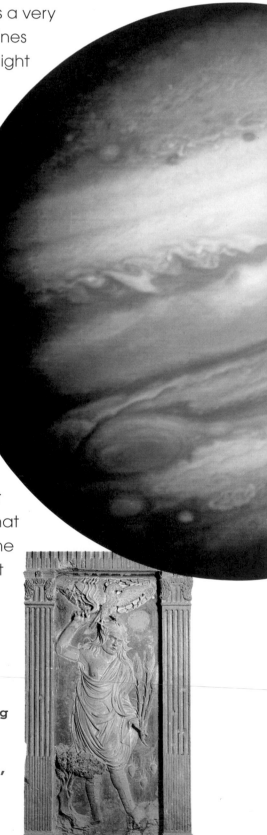

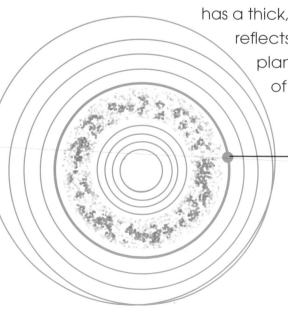

Jupiter

▷ A relief carving of the Roman god Jupiter in the Chapel of Planets, Rimini, Italy.

On Jupiter's face

When we look at Jupiter in a telescope, we see it as a disc, but it is not completely round. It is slightly flattened at the top and bottom, and bulges slightly at the sides.

The most noticeable markings on the disc are dark and light bands, parallel to one another. Astronomers call the dark bands belts and the light ones zones.

The *Voyager* picture on this page shows the typical markings seen on Jupiter's disc. All except one are features of the Jovian atmosphere. The exception is the round object (lower right). This is one of Jupiter's large moons passing by.

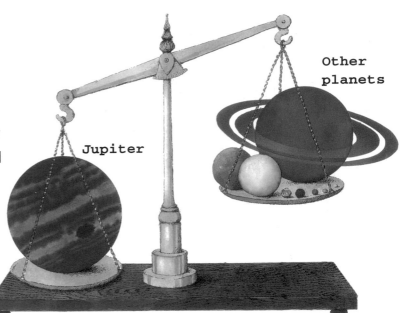

△ **Jupiter is heavier than all the other eight planets put together.**

Fast spinner

If you observe Jupiter for some time, you notice that the various markings move across the disc. They move fastest in the middle of the disc.

These observations tell us that Jupiter is a huge ball of gas that is spinning round rapidly on its axis. It is the rapid rotation that makes it flatter at the poles (top and bottom) and bulge at the equator (middle). What we see when we look at the disc is not the surface of Jupiter, but its gassy atmosphere.

By timing the movement of the markings across Jupiter's disc, we find that the planet spins round at an incredible speed. It spins round once in less than ten hours. This means that the top of the atmosphere is whizzing round at a speed of nearly 45 000 kilometres an hour.

Central heating

The temperature of most planets depends on how far away they are from the Sun. The temperature at the top of Jupiter's atmosphere is about -150°C. But this is higher than it should be for its position in the Solar System. Jupiter somehow makes its own heat. It gives off twice as much heat as it gets from the Sun.

THE JOVIAN ATMOSPHERE

Bands of coloured clouds race round Jupiter's atmosphere.

Even before space probes visited Jupiter, astronomers knew that its atmosphere contained hydrogen and hydrogen compounds. They discovered this by studying the light from Jupiter in an instrument called a spectroscope.

Space probes have revealed that hydrogen makes up more than 90 per cent of the atmosphere. The rest of the atmosphere is mostly helium. Hydrogen and helium are also the main gases found in the Sun and all the stars. They are the two most common substances in the whole Universe.

Traces of other substances, such as ammonia and water, are found in the Jovian atmosphere. There is also some methane, which is the gas we use for cooking and heating in the home.

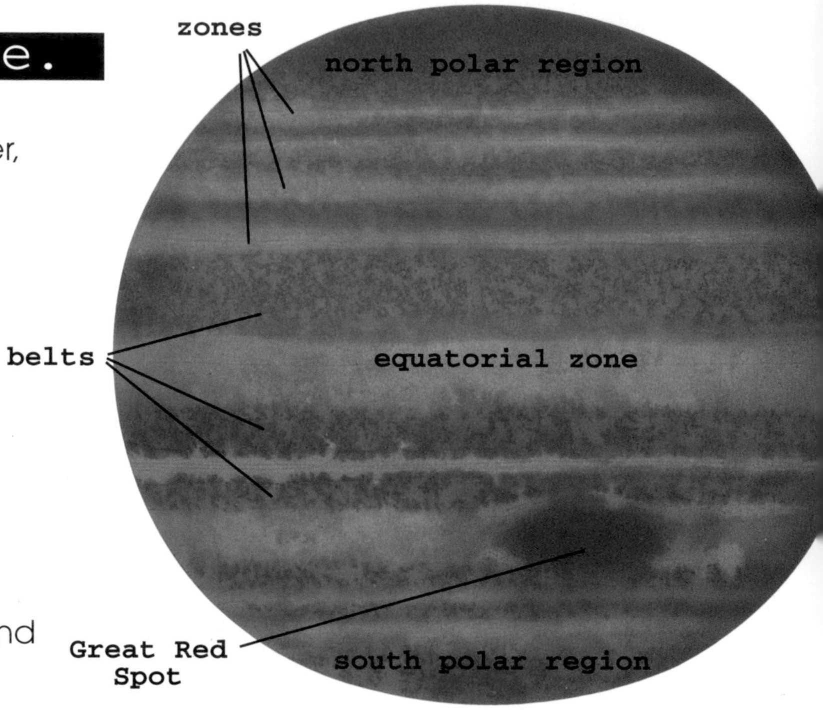

zones
north polar region
belts
equatorial zone
Great Red Spot
south polar region

△ The main features of Jupiter's atmosphere.

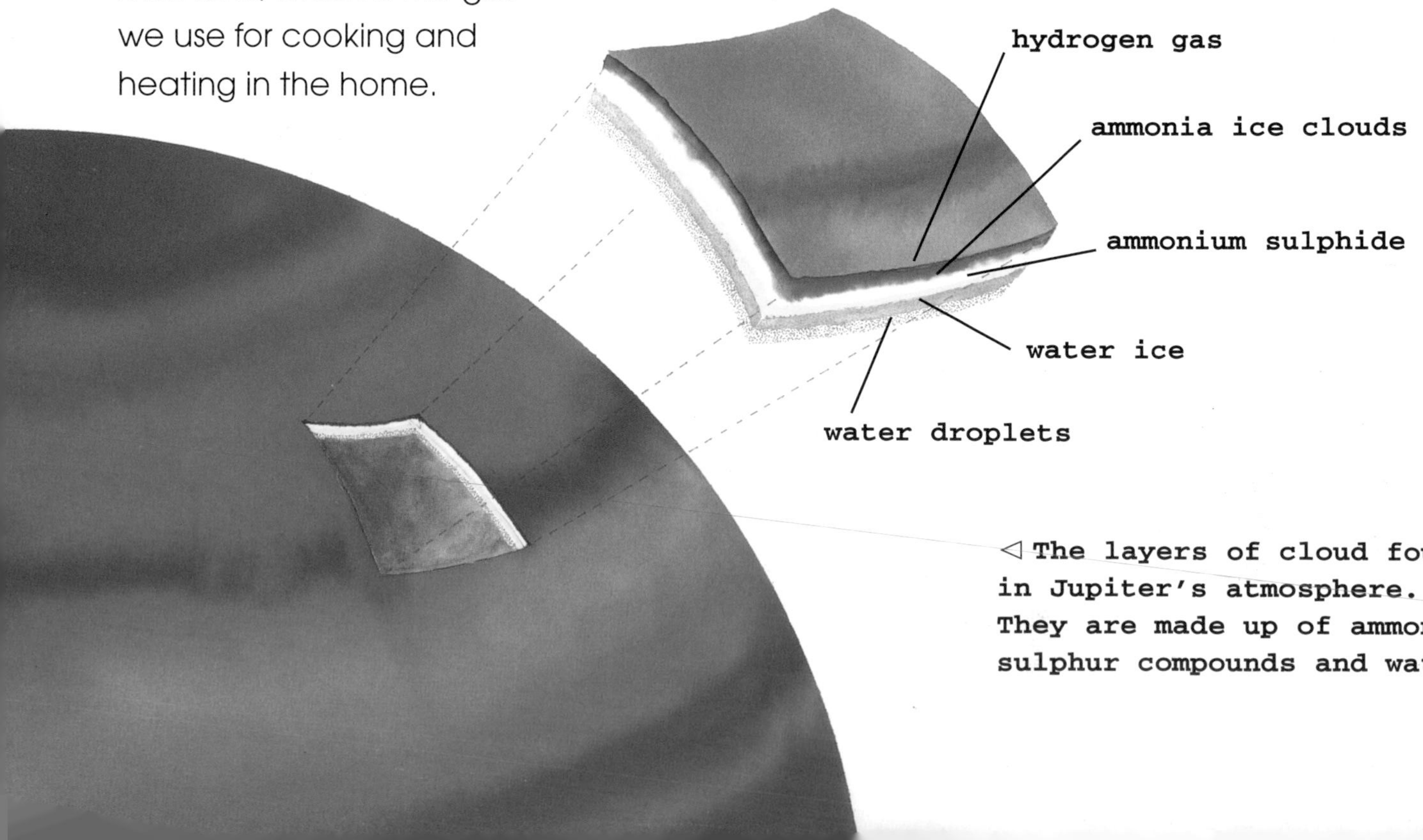

hydrogen gas
ammonia ice clouds
ammonium sulphide
water ice
water droplets

◁ The layers of cloud found in Jupiter's atmosphere. They are made up of ammonia, sulphur compounds and water.

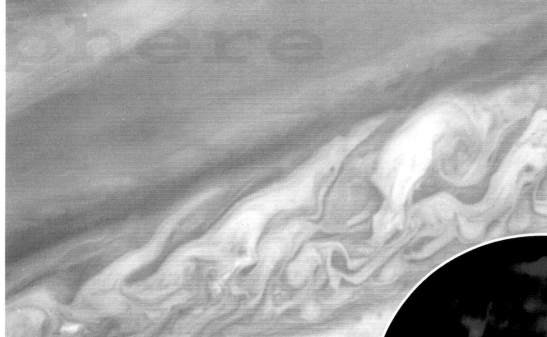

△ Swirling clouds in one of the bands in Jupiter's atmosphere make fascinating patterns.

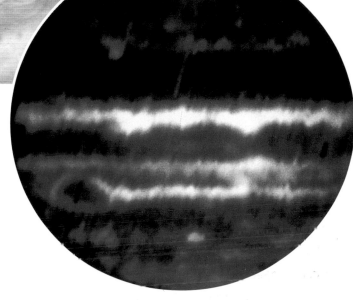

△ This infrared picture shows the heat that Jupiter gives out.

Up in the clouds

The belts (dark bands) and zones (light bands) we see in Jupiter's atmosphere through a telescope are clouds. They have been drawn out into bands parallel to the equator by the planet's rapid rotation.

The atmosphere on Jupiter is at least 1000 kilometres deep. But the clouds are found only at the very top. They appear to form three main layers. In the top layer the clouds are made up of tiny crystals of ammonia ice. Ammonia gas turns into solid crystals, or ice, when it gets very cold.

Beneath the ammonia clouds are clouds made up of a compound of ammonia and sulphur. The lowest layer is made up of clouds similar to those found on the Earth. They are formed by crystals of water ice and droplets of liquid water.

Astronomers are not sure why the cloud bands are coloured the way they are. The darker-coloured, reddish-brown bands may be lower-level clouds showing through gaps in the white ammonia cloud layer. Or the colouring may take place entirely in the ammonia layer because of chemical reactions.

Atmospheric heat

The temperature at the top of Jupiter's clouds is about –150°C. But as you go deeper down into the atmosphere the temperature quickly rises. In the bottom cloud layer, some 90 kilometres down, the temperature is probably above the freezing point of water (0°C). And it is here that Earth-type clouds of water droplets form.

Below the clouds, the temperature gradually increases the deeper you go. At the bottom of the atmosphere, where the gases turn into liquid, the temperature may be as high as high as 2000°C.

A STORMY WORLD

Fierce winds create violent storms in Jupiter's atmosphere.

Jupiter's atmosphere is in constant motion and strong winds blow parallel with the equator. It is these winds that force the clouds into parallel bands. The winds blow steadily at speeds of from 300 to 500 kilometres an hour. On the Earth, winds as strong as this are found only in hurricanes and tornadoes.

Jupiter's winds all blow parallel with the equator, but they do not all blow in the same direction. Some blow from the west, and some from the east. This is similar to what happens on the Earth, where in certain regions prevailing (usual) winds blow from the west or the east.

△ **Jupiter's Great Red Spot measures up to 40 000 kilometres across.**

▽ **The pale oval areas in Jupiter's atmosphere are storms. The round object is Europa, one of the moons of Jupiter.**

◁ Comet Shoemaker-Levy 9 scars Jupiter in July 1994.

New spots

New spots appeared in Jupiter's atmosphere in July 1994. They were not caused by storms but appeared when a comet collided with the giant planet. The comet was called Shoemaker-Levy 9 and was split into more than 20 separate pieces. As each one smashed into Jupiter, it left a spot or scar in the atmosphere.

Swirling storms

Jupiter has many more prevailing wind streams than the Earth. And they are many thousands of kilometres wide. Within the wind streams all kinds of flow patterns are set up, which we see as colourful waves and eddies.

Where wind streams moving in opposite directions meet there is a great turbulence, or swirling and churning together. In these regions, violent storms break out. We see them as swirling oval regions, usually white or red in colour. Some come and go with the years, others are more permanent.

The Great Red Spot

Astronomers spotted a particularly large red oval region on Jupiter more than 300 years ago. It is still there today, and is known as the Great Red Spot (GRS). The GRS is found just south of Jupiter's equator.

Astronomers once thought that the GRS might be the top of a huge volcano or some kind of body floating in the atmosphere. Space probes have now shown that the GRS is a huge storm centre.

△ Furious winds spiral round in the Great Red Spot.

It is like a superhurricane, with winds spiralling round and upwards in a circle. The whole Spot rotates once every six days. It rotates anticlockwise, or in the opposite direction to the hands of a clock.

The colour of the GRS varies from orange to pink and a deep brick-red. Astronomers believe that the colour is caused by phosphorus. This chemical element can be found in a red form on Earth.

RADIO JUPITER

Jupiter transmits radio signals into space.

Much of Jupiter is made up of hydrogen in the form of a liquid metal. Like all metals, this metallic hydrogen conducts, or passes on, electricity. And as the planet spins round, electric currents flow in the metallic hydrogen.

It is a basic law of science – on the Earth and everywhere else – that you always get magnetism when electric currents flow. So the electric currents flowing inside Jupiter make the planet magnetic, or give it what is called a magnetic field. A similar process inside the Earth gives our planet a magnetic field. But Jupiter's field is ten times as strong.

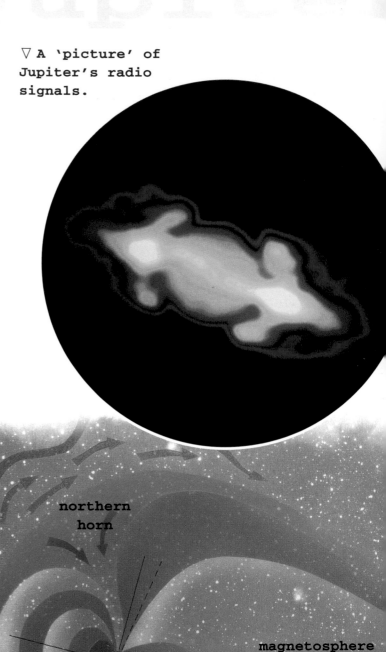

▽ A 'picture' of Jupiter's radio signals.

solar wind

northern horn

magnetosphere

southern horn

radiation belts

bow shock ——————————

magnetopause

The invisible shield

Jupiter's powerful magnetic field reaches millions of kilometres out into space. It forms a kind of magnetic 'bubble' around the planet, called the magnetosphere. But the 'bubble' is not regular in shape. It is squashed up on one side and tapers away on the other side, rather like a gigantic teardrop.

The magnetosphere is squashed up in the direction of the Sun. It gets squashed when it meets the solar wind. This is a sea of tiny particles that streams out from the Sun. These particles are like protons and electrons, which have a tiny electric charge. Jupiter's magnetosphere acts as a barrier to these particles, which the solar wind flows around.

▽ **The solar wind flows round Jupiter's magnetic field, but some particles enter through the 'horns'.**

magnetosphere tail

Jupiter's lights

The charged particles trapped in Jupiter's magnetic field sometimes shoot down into the atmosphere. This happens mainly near the north and south poles. As they hit the gas in the atmosphere, energy is given off as light. This is called the aurora. A similar thing happens on Earth. Then we call it the northern or southern lights.

Intense radiation

Some particles from the solar wind do manage to pass into the magnetic 'bubble'. They collect in two doughnut-shaped regions closer to the planet. These regions give out intense rays, or radiation, and are called radiation belts. They are similar to radiation belts found around the Earth, but much more powerful. They would be deadly to any astronauts travelling through them.

Jupiter calling

The whole magnetosphere around Jupiter is full of rapidly moving particles with enormous energy. They give off this energy as short-wave radio signals, which we can detect on the Earth.

Bursts of long-wave radio signals also occur from time to time. These long-wave signals are somehow triggered by the movement of Io, the innermost of Jupiter's four large moons.

RINGS AND MOONS

Faint rings and many moons circle round Jupiter.

Jupiter has an enormous mass and a powerful gravity, or pull, which stretches millions of kilometres out into space. With this gravity, Jupiter holds onto a large family of satellites, or moons. They include some bodies as big as planets and others that are as small as asteroids, only a few tens of metres across.

We know that Jupiter has at least 16 moons. We can see most of them from the Earth through telescopes, but some were discovered by space probes. The moons circle round Jupiter at different distances. They fall into four distinct groups of four moons each.

The inner four moons are tiny and orbit close to Jupiter. The next four moons are very large and more spaced out than the first group. Then comes a huge gap before the next four moons, which again are tiny. One, called Leda, is probably only about 10 kilometres across. The four outermost moons are many times further away still, and again are tiny. They also circle round Jupiter in the opposite direction from the other moons.

△ All Jupiter's moons look different. This is Ganymede, the largest.

◁ Jupiter's rings are too faint to be seen from Earth.

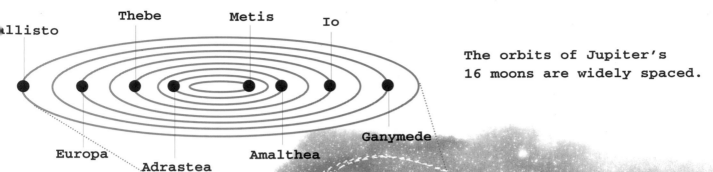

Callisto • Thebe • Metis • Io

Europa • Adrastea • Amalthea • Ganymede

The orbits of Jupiter's 16 moons are widely spaced.

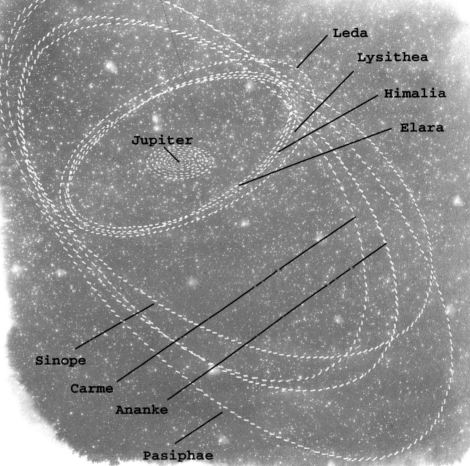

Jupiter

Leda
Lysithea
Himalia
Elara
Sinope
Carme
Ananke
Pasiphae

Captured asteroids

Most astronomers reckon that the tiny moons in the two outer groups are not true moons at all. They are probably asteroids that Jupiter captured when they strayed out of the nearby asteroid belt.

Jupiter has also captured two other groups of asteroids. They are found travelling in Jupiter's orbit in two groups, one in front of Jupiter and the other behind. They are called the Trojans.

The inner moons

By far the most important of Jupiter's moons are the 'big four' – Io, Europa, Ganymede and Callisto. These are the so-called Galilean moons, which Galileo discovered. They are so large that they can be spotted in binoculars from the Earth. We shall discuss these moons in detail later.

The two closest moons to Jupiter are the tiny Metis and Adrastea, which are only about 40 and 26 kilometres across. Discovered by the *Voyager* space probes, they both orbit about 55 000 kilometres above Jupiter's cloud tops.

The Jovian rings

In the same region as the *Voyagers* discovered Metis and Adrastea, they also spotted faint rings around Jupiter. The rings are made up mainly of fine particles and are much too faint to be seen from the Earth. The main part of the ring measures about 6000 kilometres across, with a narrower brighter section about 800 kilometres across. But fainter parts of the ring system probably reach right down to the cloud tops.

ncredible
io

INCREDIBLE IO

Io is the most amazing moon in the Solar System.

Io is very different from Jupiter's other large moons, and from all the other moons in the Solar System. Most other moons are dark and drab. But Io is as brightly coloured as a pizza, in hues of red, orange and yellow.

Most of the other moons are covered in craters, where they have been hit by meteorites. But Io has few craters, and none of them has been made by meteorites. This tells us that the moon must have a very young surface.

△ **Gas and particles shoot into Io's sky from a volcano.**

▽ **Io's colours are due to sulphur from volcanoes.**

Violent volcanoes

The reason why Io has a young surface is because volcanoes are erupting all over the place. They are pouring out new material all the time, which is covering the older landscape.

The volcanoes on Io are not pouring out molten rock, or lava, like volcanoes on the Earth do. They are pouring out mainly molten sulphur. Sulphur can exist in a variety of different forms, which vary in colour from pale yellow to deep reddish-orange. This explains why Io's surface is so vivid in colour.

What causes Io's volcanoes? Astronomers believe that it is the powerful gravity of Jupiter and the nearby moons Ganymede and Europa. As Io circles in its orbit, Jupiter and the other two moons are continually tugging at it, first this way, then that way. This sets up movements like tides inside it.

And these tides produce so much energy, as heat, that Io is mostly molten inside. Under the harder outer surface, there are lakes, maybe oceans of molten sulphur. These feed the volcanoes.

Io's plumes

When Io's volcanoes erupt, they not only pour out rivers of molten sulphur, but also shoot great fountains of particles and gases high above the surface. These fountains are called volcanic plumes. The particles are specks of sulphur, which travel a long way before they float back to the surface. The main gas is sulphur dioxide. This shoots out and cools to form tiny crystals, which fall to the ground as white snow. About a third of Io's surface is covered with sulphur dioxide snow.

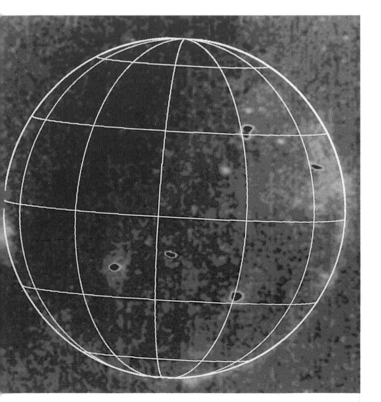

◁ **Hot spots on Io's surface show where volcanoes are erupting.**

ICY MOONS

Europa is the smallest of Jupiter's four big moons, slightly smaller than our Moon. It is very different from its neighbour Io in colour and make-up. Its surface is pale, and marked with many brownish streaks. It is very smooth – smoother than any other heavenly body we know.

Europa is so smooth because it is covered with ice. This hard icy crust may lie on top of a slushy mixture of water and ice, with rock underneath. The brown streaks on the surface appear to be cracks in the ice, which have been filled by dirty slush or rocky material welling up from below.

There are a few small recent craters on Europa, but no big old ones. So the surface must be quite young. When a meteorite hits the surface, it digs out a crater and melts the ice around it. The water probably flows back into the crater and freezes over, or water may well up from below and freeze. The crater all but disappears.

Great Ganymede

Ganymede is the largest Jovian moon, bigger even than the planet Mercury. In fact, it is the largest moon in the whole Solar System. Ganymede also has an icy crust, but it is quite different from Europa's because the ice is very dirty. The darkest areas are believed to be the oldest because they are covered with craters. The newest ones are white, showing where

△ **Callisto**

fresh ice has been exposed. One of the largest dark areas, Galileo Regio, can be seen from Earth in telescopes. Grooved, paler regions between the dark areas are younger and have fewer craters.

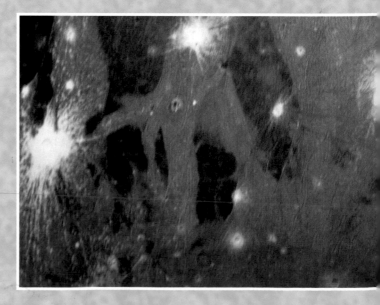

△ **Icy craters pepper Ganymede's surface.**

Cratered Callisto

Callisto is the second largest of Jupiter's moons, only slightly smaller than the planet Mercury. It has the oldest surface of the four Galilean moons and is completely covered in craters. The more recent ones are white, showing where fresh ice has been exposed.

In one place a huge meteorite hit the planet and dug out a crater nearly 3000 kilometres across, called the Valhalla Basin. Inside it, a series of rings shows where the surface was thrown up into ripples by the force of the impact. There are several other similar ringed structures on Callisto, but all of them are much smaller.

Astronomers reckon that Callisto has a hard crust of ice and rock. Underneath, it is probably like Ganymede, with a layer of water or soft ice on top of a rocky core.

Ganymede

Our Moon

△ With a diameter of 5276 kilometres, Ganymede is much bigger than our own Moon.

▷ Europa's dark markings prove to be troughs and ridges where the surface has moved and fractured.

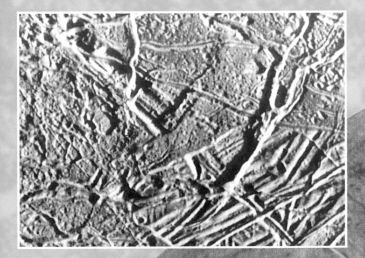

▷ Europa's surface is criss-crossed with dark markings.

THE RINGED WONDER

the ringed wonder

Shining rings make Saturn a beautiful sight in a telescope.

△ **The symbol for Saturn represents a sickle. Saturn was the god of the harvest.**

Saturn is second in size only to Jupiter among the planets of the Solar System. Even though it sometimes travels more than 1500 million kilometres from the Earth, it is big enough to be visible in the night sky. But it is the faintest of the five planets we can see with the naked eye. It would be even fainter if it did not have its shining rings.

Ancient astronomers were familiar with the ringed planet. The ancient Romans named it after Saturn, who was god of the harvest and first ruler of the other gods. He was the father of many other gods, including Jupiter.

Through the telescope

Even a small telescope reveals that there is something unusual about Saturn. The planet shows up, not as a disc like Jupiter, but as a more oval image. A bigger telescope will show what causes this: a set of rings around the planet. These rings circle the planet's equator. We can see three main rings, named A, B and C from the outside going towards the planet. Ring B is the brightest of the three and ring C the faintest. There is a noticeable gap between the A and B rings, called the Cassini division.

▷ **A *Voyager* picture of Saturn and some of its moons.**

▽ **The god Saturn, in the Chapel of Planets, Rimini, Italy.**

▽ **Saturn is the sixth planet from the Sun, between Jupiter and Uranus.**

Saturn

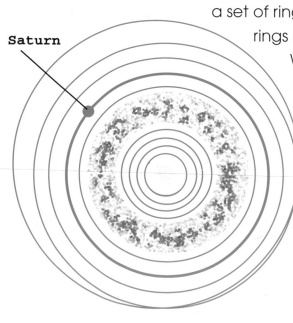

Belts and zones

On the main body, or disc, of Saturn dark and light bands show up. They are like the belts and zones we see on Jupiter's disc, but very much fainter. They, too, are parallel bands of clouds in Saturn's deep atmosphere.

Spinning round

Saturn spins round in space on its axis nearly as fast as Jupiter, in just over 10½ hours. Because the planet is made up mainly of gas, the fast spinning makes it flatten at the poles and bulge out at the equator. This effect is more noticeable with Saturn than with Jupiter.

Saturn does not spin round in an upright position as it circles in orbit around the Sun. Like the Earth, it spins round an axis that is tilted at an angle. Because of this tilt, we see different views of the rings during the 30 years Saturn takes to circle the Sun.

Floating on water

Saturn is even lighter for its size, or is less dense, than Jupiter. Its density is less than the density of water. This means that if you had a big enough bowl and filled it with water, Saturn would float. Jupiter and every other planet would sink.

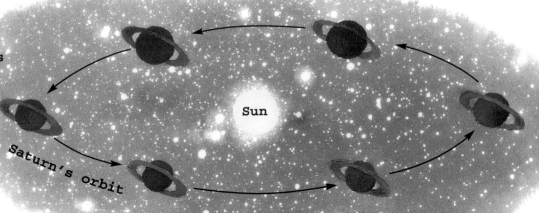

▷ Saturn's rings always point in the same direction in space.

Sun

Saturn's orbit

THE WINDY ATMOSPHERE

the windy atmosphere

Fierce winds blow in Saturn's atmosphere.

Saturn has much the same structure as Jupiter. Its atmosphere is made up of hydrogen and helium, and underneath there is a deep ocean of liquid hydrogen. But there is not as much helium in Saturn's atmosphere as there is in Jupiter's.

Scientists think that some of this helium falls out of the atmosphere into the ocean as rain. When the helium rain falls into the ocean, it gives up energy as heat. Over billions of years, this has heated up the whole planet. This could explain why the planet gives off more than twice as much heat as it receives from the Sun.

△ A natural-colour picture of Saturn.

▽ A false-colour picture of Saturn shows the bands in the atmosphere more clearly.

Clouds and winds

Clouds form in Saturn's atmosphere and are drawn out into bands by strong winds blowing parallel with the equator. This is exactly what happens with Jupiter. And, as in Jupiter's atmosphere, clouds of ammonia crystals, sulphur compounds and water form at different levels.

The winds in Saturn's atmosphere blow in different directions, some towards the east and some towards the west. Saturn's winds are very strong. In some regions they blow at speeds of more than 1800 kilometres an hour. This is much stronger than any winds on the Earth.

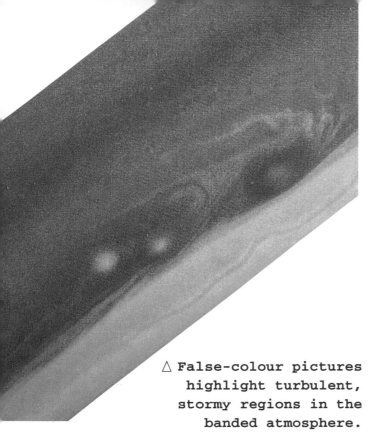

△ **False-colour pictures highlight turbulent, stormy regions in the banded atmosphere.**

Stormy weather

When furious winds travelling in opposite directions meet, the atmosphere gets churned up. Huge storms break out, as gigantic whirlwinds spin round at fantastic speeds. Thunder booms and lightning flashes as in thunderstorms on the Earth. But on Saturn such storms are much more intense. They cover thousands of kilometres and last for months, even years, at a time.

These storms show up in pictures taken by spacecraft as round or oval spots that are white or red in colour. They are similar to the spots on Jupiter, but not so marked.

▷ **Voyager 2 spotted this curious curled cloud on the edge of a bright white cloud band in Saturn's atmosphere.**

Magnetic Saturn

Like Jupiter, Saturn has under its liquid hydrogen ocean a deep layer of hydrogen in the form of a liquid metal. And electric currents in this layer make Saturn magnetic. Its magnetism is not as powerful as Jupiter's, but it is much more powerful than the Earth's.

So Saturn, too, is surrounded by a magnetic 'bubble', or magnetosphere. And when electrically charged particles from the magnetosphere hit Saturn's atmosphere, they create the light displays we call the aurora. Particles whizzing round in the magnetosphere also set up radio waves, as they do on Jupiter. But Saturn is a very weak 'radio station' compared with Jupiter.

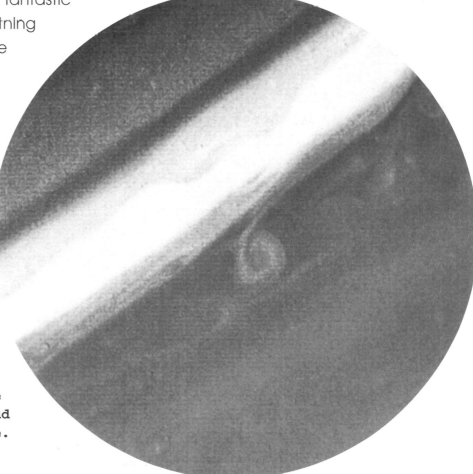

THE GLORIOUS RINGS

Saturn's rings
dazzle the eye.

In a telescope, we can see three rings around Saturn – A, B and C. The outer edge of the A ring is about 76 000 kilometres above the planet's atmosphere. This means that the visible ring system of Saturn measures about 272 000 kilometres across.

The three rings are all different. The inner C ring, also called the crêpe ring, is faint. It is also quite transparent. We can see the body of the planet showing through it. The B ring is the widest ring of the three and by far the brightest. It also casts the darkest shadow on the body of the planet.

Saturn's rings are wide, with the B ring being the widest at about 25 000 kilometres across. But they are very thin. We know this because at certain times the rings appear edge-on when we view them from Earth. And then they almost disappear from view.

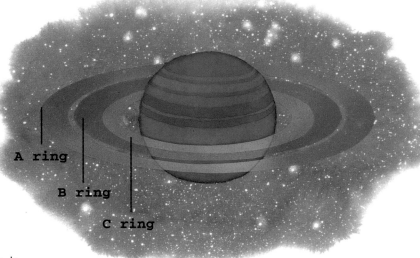

A ring

B ring

C ring

△ **Three rings of Saturn can be seen from the Earth – A, B and C.**

This happened last in 1995, and will happen again in 2009. Astronomers think that on average the rings are only about 30 metres deep.

Encke
division

Cassini
division

△ **The A, B and C rings show clearly here. So do the 'gaps' or divisions in the rings.**

▷ When viewed from
the unlit side,
Saturn's B ring
looks darkest.

Gaps in the rings

Between the B ring and the A ring is a gap where there appears to be no rings. This is called the Cassini division after the Italian astronomer Jean Dominique Cassini, who first spotted it in the 1600s. There is also a smaller gap in the A ring itself. This is called the Encke division, after German astronomer Johann Encke, who discovered it in the 1830s.

No one is certain why the gaps occur. But it is probably something to do with forces set up by the gravity of Saturn's moon Mimas, which orbits just outside the rings.

The invisible rings

Space probes have shown that there are several other rings around Saturn, besides the A, B and C. But they are too faint to be visible from Earth. Inside the C ring is a D ring, which extends right down to the planet's cloud tops. And there are three rings beyond the A ring: a narrow and curiously twisted F ring, a fainter G ring and a very faint but broad E ring. The E ring may reach out to more than 400 000 kilometres above Saturn's clouds.

▷ The Cassini division is
not empty but contains
quite a few ringlets.

Origins

No one is sure where Saturn's rings came from. They might be made up of material left over after Saturn formed. Or they could be the remains of a moon. Saturn's strong gravity might have pulled this moon closer and closer. When it got too close, Saturn's powerful gravity would have torn it into little pieces, which in time formed into rings.

△ Saturn's rings are made up of millions of icy lumps.

RINGS AND RINGLETS

Saturn's rings are made up of thousands of separate ringlets.

From the Earth, Saturn's rings look as if they are solid sheets of material. But for a long time, astronomers have known that this is not possible. Solid rings would immediately be pulled apart and shattered into tiny bits by Saturn's powerful gravity.

The rings are actually made up of bits, or particles, of material whizzing round in orbit at high speed. Because the particles travel fast, they appear blurred when we look at them. And this blurring makes them appear to form a continuous sheet.

▽ Our view of the rings changes as time goes by.

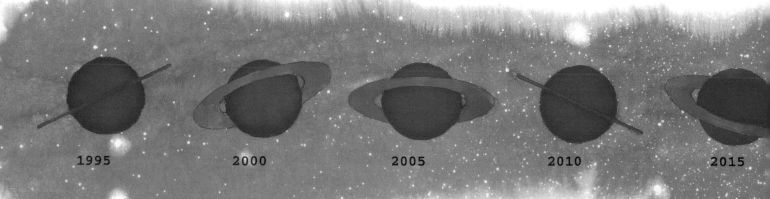

1995　　2000　　2005　　2010　　2015

Saturn's snowballs

The particles in Saturn's rings seem to be rather like snowballs. They are made up mainly of ice, together with some rocky material. They vary in size from specks of dust to chunks up to 10 metres across.

The particles in the rings travel round the planet at different speeds. The ones in the rings closest to the planet travel fastest. Particles in the C ring take six to eight hours to circle Saturn, while those in the distant A ring take twice as long.

Racing ringlets

Each of Saturn's rings looks like a continuous sheet in a telescope. But close-up photographs show that each ring is in fact made up of thousands of tiny ringlets. They make the ring system look rather like an old record disc with its many grooves.

The groovy appearance of the rings is thought to be caused by gravity forces acting on the ring particles. In some regions, forces cause the particles to clump together in places, leaving gaps in between. As each clump travels round, it forms a separate ringlet. In other regions, gravity forces make the ring particles form waves, rather like water waves, and the ring literally does become groovy.

The disappearing spokes

Close-up photographs of the rings show up curious features, such as dark spokes in the B ring. These spokes cut across the ring, fanning out rather like the spokes of a bicycle wheel. They appear suddenly and grow thousands of kilometres long in minutes. They rotate with the rings for several hours and then break up. Astronomers think that the spokes are caused by clouds of dust.

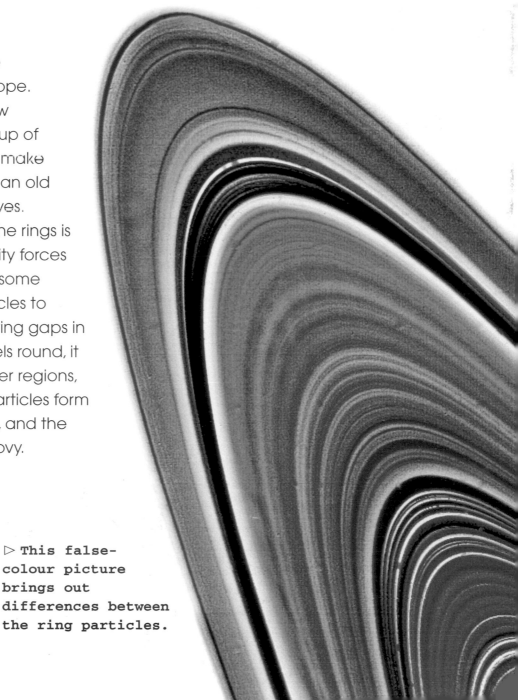

2020

▷ **This false-colour picture brings out differences between the ring particles.**

SATURN'S SATELLITES

Tiny 'shepherd' moons keep Saturn's rings in place.

As far as we know, Saturn has more satellites, or moons, circling around it than any other planet in the Solar System. There are at least 23. Through a telescope, only ten satellites can be seen. The others were discovered by space probes.

Saturn's moons vary widely in size from tiny bodies just a few kilometres across to a planet-sized body that is bigger than Mercury. They can be divided into groups according to their different orbits. Three moons are found near the edge of the ring system. They were discovered by the

Voyager probes. Next comes a group of much larger moons that have been known for centuries: Mimas, Enceladus, Tethys, Dione and Rhea.

Further out is a pair of moons relatively close together, Titan and Hyperion. Iapetus orbits twice as far away. Finally there is remote Phoebe, circling nearly 13 million kilometres from Saturn. It also circles in the opposite direction from the other moons. So astronomers think that it is probably a wandering asteroid that Saturn captured a long time ago.

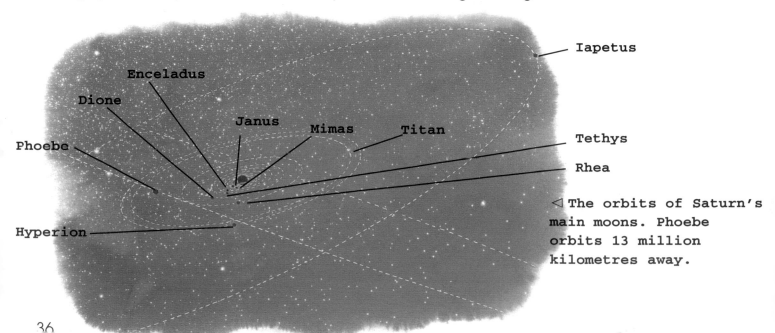

Iapetus

Enceladus

Dione

Janus

Mimas

Titan

Phoebe

Tethys

Rhea

Hyperion

◁ **The orbits of Saturn's main moons. Phoebe orbits 13 million kilometres away.**

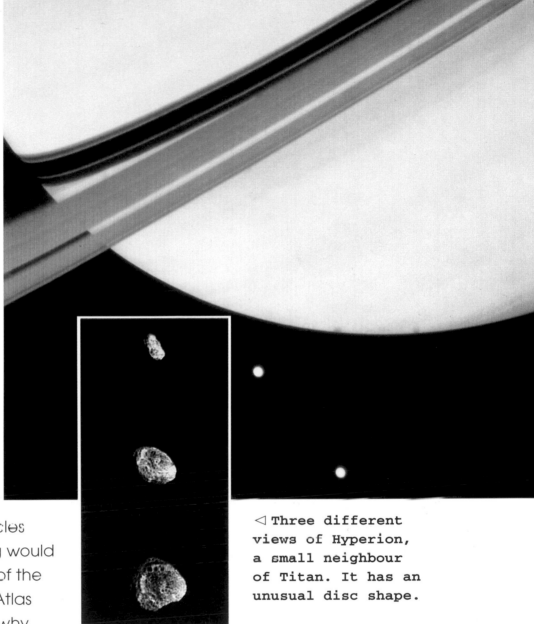

▷ Saturn and two of its moons – Dione (top) and Tethys. Tethys is casting a shadow on the planet.

The shepherding moons

Atlas is the innermost moon of Saturn. It circles just beyond the outer edge of the A ring. Astronomers believe that it keeps the particles of the ring in place. The particles in the outer part of the ring would naturally tend to stray out of the ring, but they do not. And Atlas appears to be the reason why. Although Atlas is less than 40 kilometres across, its gravity seems to keep the A ring particles in place – just like a shepherd keeps a flock of sheep together. For this reason, Atlas is called a shepherd moon.

Two larger shepherd moons keep the particles in the narrow F ring in place. There is one on either side: Prometheus (inner) and Pandora (outer). As they travel in their orbits at slightly different speeds, they keep the particles between them circling in a narrow band.

◁ Three different views of Hyperion, a small neighbour of Titan. It has an unusual disc shape.

Sharing orbits

Some of the tiny new moons discovered by the *Voyager* space probes travel in the same orbits as other moons. They are known as co-orbitals. For example, two co-orbitals travel with Tethys in its orbit. One travels in front of Tethys, one behind, and always at the same distance. These moons are called Telesto and Calypso.

SATURN'S MAJOR MOONS

The larger moons of Saturn are icy worlds.

Among the many moons of Saturn, Titan is in a class of its own because of its huge size (see page 40). It is quite unlike the seven other major moons, which range in size from about 400 to 1500 kilometres across. These typical moons are made up mainly of water ice, and contain only a little rocky material.

Mimas is the smallest of these moons, and is the one closest to Saturn. It is covered with craters, one of which is huge (about 130 kilometres across).

Mimas's neighbour Enceladus is larger and looks quite different. It has fewer craters and some regions with no craters at all. Processes must be at work that in time cover up any craters that form. Water 'volcanoes' might spurt out of the surface, cover the landscape and freeze into fresh ice. Fresh ice reflects light like a mirror. This could explain why the moon reflects light better than any other body known.

△ Dione, one of Saturn's moons. The largest crater is 60 kilometres wide.

Tethys, Dione and Rhea

The next three moons going away from Saturn are more than twice the size of Enceladus. They are Tethys, Dione and Rhea. Tethys is another moon that is heavily cratered. One of the craters, named Odysseus, measures about 400 kilometres across. The other main feature on Tethys is a huge valley and canyon system called Ithaca Chasma. It extends for 2000 kilometres, or almost three-quarters of the way around the moon.

◁ Iapetus is bright on one side and very dark on the other.

▽ Icy Enceladus reflects light better than any other body known.

Tethys shares its orbit with two tiny moons, the co-orbitals Telesto and Calypso. Dione, the next moon out, also shares its orbit, with at least one moon (Helene). It is a near twin of Tethys in size and is also well cratered. Bright wispy streaks cover its surface in many places. They are thought to be regions where fresh ice has seeped out onto the surface from the moon's interior.

Rhea and Iapetus

The largest of Saturn's icy moons are Rhea and Iapetus. They are much the same size (about 1500 kilometres across), but are far apart. Rhea is a member of the inner group of moons and orbits about 530 000 kilometres from Saturn. Iapetus orbits nearly 3 000 000 kilometres further out.

The two moons are quite different to look at. Rhea has a bright, heavily cratered surface that looks rather like the highlands of our Moon. Iapetus is bright and well cratered in places, but also has a huge region that is as black as tar.

TANTALIZING TITAN

Titan has a thicker atmosphere than the Earth.

Saturn's largest moon Titan is the second biggest moon in the whole Solar System. Measuring some 5150 kilometres across, it is only fractionally smaller than Jupiter's moon Ganymede. And it is thought to be like Ganymede in make-up. Beneath a hard icy crust there is probably a deep layer of water or a slushy mixture of water and ice. Underneath this is a rocky core.

Titan is quite different from Ganymede in one respect – it has a thick atmosphere. In fact, it is the only moon in the Solar System to have one.

Close-up photographs of Titan taken by the *Voyager* space probes show that the atmosphere is coloured orange and is full of haze and clouds. The clouds are so thick that we cannot see through them. So we do not know what Titan's surface is really like. We may find out when the *Huygens* probe lands on Titan in 2004.

▽ **Layers of haze (shown in blue here) are found high up in Titan's atmosphere.**

▽ **Thick clouds cover Titan.**

The cloudy atmosphere

The main gas in Titan's atmosphere is nitrogen, which is the main gas in the Earth's atmosphere. There are also small amounts of helium, and traces of many other gases. The most important one of these is methane, which is the main gas in the natural gas we find on Earth. Methane is one of several hydrocarbon (hydrogen and carbon) compounds in Titan's atmosphere. There are also compounds of carbon and nitrogen. Scientists reckon that compounds like these were probably present in the Earth's first atmosphere billions of years ago.

Surprisingly, there is more gas in Titan's atmosphere than there is in the Earth's. And the pressure at the surface of Titan is more than one-and-a-half times the normal atmospheric pressure on Earth.

Scientists think that the thick orange haze around Titan is caused by a fine mist of carbon compounds. Further down there are probably darker clouds made up of tiny crystals of frozen methane. Out of these clouds liquid methane may fall to the surface as drizzly rain.

△ **The date is October 2004. The *Huygens* probe is parachuting down to Titan's surface.**

On the surface

The temperature of Titan's surface is about -180°C. This is close to the temperature at which methane can exist as a liquid or solid (ice). This means that it might play the same part in Titan's atmosphere as water plays in the Earth's.

So on Titan, there could be lakes and rivers of liquid methane running through a landscape made up of methane ice. It could rain or snow methane depending on the temperature.

▽ **Light and dark areas on Titan's surface, spotted by the Hubble Space Telescope.**

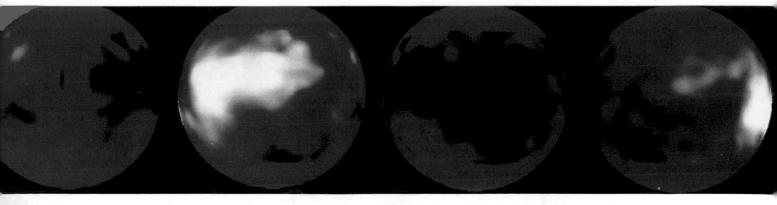

TIME LINE

650 BC
First observations of Saturn recorded about this time in Mesopotamia, in the Middle East.

AD 1514
The Polish priest-astronomer Nicolaus Copernicus observes Saturn in the constellation Scorpius.

1563
The Danish astronomer Tycho Brahe records a conjunction (appearing close together in the sky) of Saturn and Jupiter.

1610
The Italian astronomer Galileo becomes the first person to observe Jupiter in a telescope, and spies the four largest of the planet's moons – Io, Europa, Ganymede and Callisto. They become known as the Galilean moons. He also observes that Saturn has 'ears' each side, the first indication of the planet's rings.

1655
The Dutch astronomer Christiaan Huygens discovers Saturn's largest moon Titan.

1656
From his observations of Saturn of the previous year, Huygens suggests that Saturn has rings around it.

1664
The English physicist Robert Hooke makes the first record of Jupiter's Great Red Spot.

1675
The Italian astronomer Giovanni Cassini discovers a gap in Saturn's ring system, which becomes known as the Cassini division.

1690
Cassini discovers that different parts of Jupiter rotate at different speeds. This suggests that we see the planet's gaseous atmosphere, not a solid surface.

1705
Jacques Cassini, son of Giovanni, suggests that Saturn's rings are not solid, as was then thought.

1837
The German astronomer Johann Encke discovers a gap in Saturn's A ring, which becomes known as the Encke division.

1850
The American astronomer William Bond discovers the inner, C ring of Saturn, now known as the crêpe ring.

1866
The American astronomer Daniel Kirkwood suggests that the gaps in Saturn's rings are caused by forces set up by the gravity of Saturn's inner moons.

1875
The Scottish physicist James Clerk Maxwell proves theoretically that Saturn's rings cannot be solid, because they would be broken up by the planet's powerful gravity.

1932

Astronomers detect ammonia and methane in Jupiter's atmosphere.

1933

The English comedian Will Hay, also a noted amateur astronomer, discovers a prominent white spot on Saturn.

1955

Radio signals are detected coming from Jupiter, by B. Burke and F. Franklin in the United States.

1963

Astronomers suggest that Jupiter's Great Red Spot might be a wave feature in the atmosphere at the top of a tall mountain.

1972

Radar reflections from Saturn's rings suggest that they are made up of icy particles.

1973

In December, the *Pioneer 10* probe takes the first close-up photographs of Jupiter and shows that the Great Red Spot is a huge hurricane-like feature.

1979

In March, *Voyager 1* takes detailed close-up photographs of Jupiter and many of its moons. It discovers rings around the planet, several new moons, and spies volcanoes erupting on the Galilean moon Io. *Voyager 2* follows in July. In September, *Pioneer-Saturn* (formerly *Pioneer 11*), takes the first close-up photographs of Saturn and discovers a new ring.

1981

Voyager 2 encounters Saturn, taking close-up photographs of the planet and many of its moons. It discovers several new moons and pictures new rings, some flanked by tiny 'shepherd' moons, which 'herd' the ring particles.

1990

The Hubble Space Telescope is launched and begins a regular check on Jupiter and Saturn and their moons.

1994

In July, the many pieces of Comet Shoemaker-Levy 9 crash into Jupiter, creating spectacular fireballs and leaving behind 'scars' in the atmosphere.

1995

In August, Saturn's moons are seen edge-on, and nearly disappear from view. In December, the *Galileo* probe reaches Jupiter after a six-year journey (via Venus, Earth and the asteroid belt, where it photographed the asteroids Gaspra and Ida). *Galileo* drops a smaller probe into Jupiter's atmosphere before going into orbit.

1997

The *Cassini-Huygens* probe is launched on a seven-year roundabout journey (via Venus, Earth and Jupiter) to Saturn.

2004

Cassini aims to go into orbit around Saturn and drop the *Huygens* landing probe down to Titan to find out about the moon's surface and atmosphere.

Jupiter data

Diameter at equator:	142 200 km
Volume:	1320 times Earth's volume
Mass:	318 times Earth's mass
Density:	1.3 times density of water
Gravity at surface:	2.6 times Earth's gravity
Distance from Sun average:	778 000 000 km
furthest:	816 000 000 km
closest:	741 000 000 km
Spins on axis in:	9 hours 55 minutes
Circles round Sun in:	11.9 years
Speed in orbit:	47 000 km an hour
Temperature:	-150°C
Moons:	16+

Saturn data

Diameter at equator:	120 000 km
Volume:	744 times Earth's volume
Mass:	95 times Earth's mass
Density:	0.7 times density of water
Gravity at surface:	1.2 times Earth's gravity
Distance from Earth average:	1 427 000 000 km
furthest:	1 507 000 000 km
closest:	1 347 000 000 km
Spins on axis in:	10 hours 40 minutes
Circles round Sun in:	29.5 years
Speed in orbit:	35 000 km an hour
Temperature:	-180°C
Moons:	18+

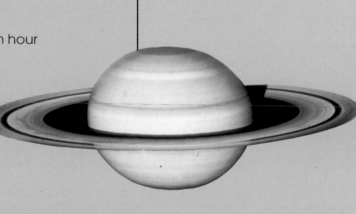

Jupiter notes

ONE A YEAR

Like all the other planets, Jupiter travels through a narrow band of the heavens called the zodiac. In the zodiac there are 12 constellations, and Jupiter takes nearly 12 years to travel once around the heavens as it circles the Sun. So on average, it passes through a different constellation every year. For example, in mid-2000 Jupiter was in the constellation Taurus. In 2002 it will be in Cancer.

SPEEDY MOVER

Jupiter spins round in space faster than any other planet. It takes less than ten hours for this giant to spin round once. This means that a point on its surface near the equator travels at a speed of nearly 45 000 km. This is about 30 times faster than the Earth spins.

WHAT A SPOT

The Great Red Spot, the huge storm that rages on Jupiter, has been observed for more than three centuries. The English scientist Robert Hooke spotted it in 1664. Sometimes it has disappeared from view, most recently in 1976. But it always returns. It covers a vast area, up to about 40 000 km long and 28 000 km wide. Three Earths could fit into it side by side.

NEARLY A STAR

In some respects Jupiter is more like a star than a planet like Earth. For one thing it is very much bigger than the average planet – bigger than all the other planets put together. It is made up mainly of hydrogen and helium – just like a star. And it generates its own heat – like a star. If it had become very much bigger, maybe it would have turned into a star.

MORE JUPITERS

Astronomers have been searching the heavens for planets around other stars for a long time. And over the past few years, they have found some. They have worked out that most of these new planets are as big as Jupiter or even bigger.

LIFE ON JUPITER

Jupiter is cold on the outside, but warms up as you descend through the atmosphere. Some scientists have suggested that somewhere in the atmosphere, some forms of life might exist, feeding on the cocktail of chemicals the atmosphere contains. These life forms would stay forever floating in the atmosphere.

VOLCANIC PLUMES

Galileo's moon Io is remarkable for its many active volcanoes, which pour out sulphur when they erupt. They also shoot particles and gases high into the sky, forming what is called a volcanic plume. One volcano , has been observed to shoot material nearly 300 km high.

Saturn notes

ICY RINGS

The rings of Saturn span a distance of over 250 000 km. They show up brilliantly because they are composed mainly of ice. But altogether it is estimated that there is little more ice in the rings than there is in an average comet, or even in the polar ice caps of our own world.

FILLING THE GAPS

In telescopes, two dark gaps show up in Saturn's rings – one in the A ring, called the Encke division, and a larger one between the A and the B ring, called the Cassini division. When the *Voyager* space probes visited the planet, they found that the gaps were not empty after all but contained many tiny ringlets.

SATURN'S SEASONS

Like all the planets, Saturn spins round in space. In fact, it spins round faster than any other planet except Jupiter. The axis around which it spins is tilted in relation to the path it follows through the heavens. This means that it is tilted more towards the Sun at some times than at others. The Earth's axis is tilted in the same way, which is what brings about the changes in temperature with the seasons. So Saturn experiences seasonal changes in temperature, too. These changes have an effect on the winds in the atmosphere

GLOSSARY

asteroids
Lumps of rock that orbit the Sun in a band between the orbits of Mars and Jupiter.

astronomy
The scientific study of the heavenly bodies.

atmosphere
A layer of gases around a planet or a star.

atmospheric pressure
The weight of the atmosphere pressing down.

aurora
A glow in the sky in polar regions produced when particles enter the atmosphere. Aurorae on Earth are known as the northern and southern lights.

axis
An imaginary line around which a body spins.

belts
The dark bands seen in the atmosphere of Jupiter and Saturn.

chemical elements
The basic chemicals found in all kinds of substances; the building blocks of matter.

comet
An icy ball of matter that starts to glow when it nears the Sun.

constellation
A group of bright stars that appear in the same part of the sky.

core
The centre part of a planet or moon.

corona
A circular feature found on Venus.

crater
A pit dug out of the surface of a planet or moon, caused by a meteorite.

day
The time the Earth takes to spin round once in space. A planet's day is the time the planet takes to spin round once on its axis.

disc
The circular appearance of a planet as we see it in a telescope.

elements see chemical elements

equator
An imaginary line around the Earth, midway between the North and South Poles.

galaxy
A great star 'island' in space, containing many billions of stars.

Galilean moons
The four main moons of Jupiter, first seen by Galileo.

gravity
The force that attracts one lump of matter to another.

Great Red Spot
A huge storm that rages on Jupiter.

hydrogen
The main substance found in Jupiter and Saturn. It is found as a gas in the atmosphere, as a liquid in the great ocean that covers each planet, and as a kind of liquid metal in a layer beneath the ocean.

Jove
Another term for Jupiter. Jovian means to do with Jupiter, such as the Jovian atmosphere.

magnetosphere
The region in space around a planet in which its magnetism acts.

meteorite
A piece of rock or metal from outer space that falls to the ground.

moon
A natural satellite of a planet.

orbit
The path in space a heavenly body follows when it circles around another.

planet
A large body that circles in space around the Sun.

radiation belts
Regions of intense radiation around a planet, containing electrically charged particles that have been captured by the planet's magnetic field.

ringlet
A very narrow ring.

ring system
The set of rings around a planet. Jupiter, Saturn, Uranus and Neptune all have ring systems.

satellite
A body or an object that circles around another body in space. The Moon is a natural satellite of the Earth. The Earth also has many artificial satellites.

shepherd moon
A tiny moon found near the edge of a planet's ring that somehow keeps the ring particles in place.

solar
To do with the Sun.

Solar System
The Sun's family, which includes the planets and their moons, asteroids and comets.

solar wind
A stream of electrically charged particles given out by the Sun.

space probe
A spacecraft that escapes the Earth's gravity and travels to other planets and their moons.

spokes
Curious spoke-like features observed in Saturn's B ring, thought to be caused by rising sheets of dust.

star
A huge ball of hot gases, which gives off energy as light, heat and other radiation.

terrestrial
Like the Earth.

Trojans
Groups of asteroids that have been captured by Jupiter and circle the Sun in Jupiter's orbit.

turbulence
Churning effects in a planet's atmosphere due to swirling currents.

Universe
Everything that exists – the Earth, space, stars, planets and all the other heavenly bodies.

volcano
A place where molten rock forces its way to the surface of a planet or moon.

year
The time the Earth takes to circle once in space around the Sun. A planet's year is the time it takes the planet to circle round the Sun.

zodiac
An imaginary band in the heavens in which the Sun and the planets are found.

zones
The light-coloured bands seen in the atmospheres of Jupiter and Saturn.

INDEX